a cap in the sun

Making Shade

Elizabeth Austen

Smithsonian

a hand in the sun

a hat in the sun

a parasol in the sun

an umbrella in the sun

a tent in the sun

a porch in the sun

us in the shade

STEAM CHALLENGE

The Problem

It is a hot, hot day. You need some shade! What can you make to give you shade?

The Goals

- Make a cover to give you shade.
- The cover can be made from any materials you like.
- It should shade your whole body.

1 Research and Brainstorm

Learn about shade.

2 Design and Build

Draw your plan. Build your shade cover!

3 Test and Improve

Stand under your shade cover. Then, try to make it better.

4 Reflect and Share

What did you learn?

Consultants

Amy Zoque
STEM Coordinator and Instructional Coach
Vineyard STEAM School
Ontario Montclair District

Siobhan Simmon
Marblehead Elementary
Capistrano Unified School District

Publishing Credits

Rachelle Cracchiolo, M.S.Ed., *Publisher*
Conni Medina, M.A.Ed., *Editor in Chief*
Diana Kenney, M.A.Ed., NBCT, *Series Developer*
Emily R. Smith, M.A.Ed., *Content Director*
Véronique Bos, *Creative Director*
Robin Erickson, *Art Director*
Stephanie Bernard, *Associate Editor*
Mindy Duits, *Senior Graphic Designer*
Smithsonian Science Education Center

Image Credits: all images from iStock and/or Shutterstock.

Library of Congress Cataloging-in-Publication Data

Names: Austen, Elizabeth, author. | Smithsonian Institution.
Title: Making shade / Elizabeth Austen ; Smithsonian.
Description: Huntington Beach, CA : Teacher Created Materials, Inc., [2020] | "Smithsonian Institution"--copyright statement. | Audience: K to grade 3. | Description based on print version record and CIP data provided by publisher; resource not viewed.
Identifiers: LCCN 2018055270 (print) | LCCN 2018058418 (ebook) | ISBN 9781425859893 (eBook) | ISBN 9781493866441 (pbk.)
Subjects: LCSH: Shades and shadows--Juvenile literature. | Sun--Juvenile literature.
Classification: LCC QC381.6 (ebook) | LCC QC381.6 .A9275 2020 (print) | DDC 535.4--dc23
LC record available at https://lccn.loc.gov/2018055270

Smithsonian

Teacher Created Materials

5301 Oceanus Drive
Huntington Beach, CA 92649-1030
www.tcmpub.com

ISBN 978-1-4938-6644-1

Printed in China 51497